JED

ME AND MY VEGGIES
WRITTEN AND ILLUSTRATED BY
ISAAC WHITLATCH

ME AND MY
WRITTEN AND ILLUSTRATED BY

VEGGIES

ISAAC WHITLATCH

LANDMARK EDITIONS, INC.

1402 Kansas Avenue • Kansas City, Missouri 64127
(816) 241-4919

Dedicated to

My brother, Luke,
for all the veggies he's eaten for me;
My parents,
for their help and inspiration;
My wild and crazy friends —
Ben, Marc, Bret, Matt, Steve,
Danny, Chris, Joe, Fred, Jason
Robby, Todd, Nathan and Randy;

And to garbage disposals everywhere.

Fourth Printing

International Standard Book Number: 0-933849-09-5

International Standard Book Number: 0-933849-16-8 (LIB.BDG.)

Library of Congress Cataloging-in-Publication Data
Whitlatch, Isaac, 1974-
Me and My Veggies.
Summary: The author relates his dislike of vegetables and reveals secret tactics used to survive the ordeal of eating them.

[1. Vegetables—Fiction. 2. Food habits—Fiction.
3. Children's writings.]

I. Title.
PZ7.W5914Me 1987 [E] 87-2920

Editorial Coordinator: Nancy R. Thatch
Creative Coordinator: David Melton

Printed in the United States of America.

ME AND MY VEGGIES

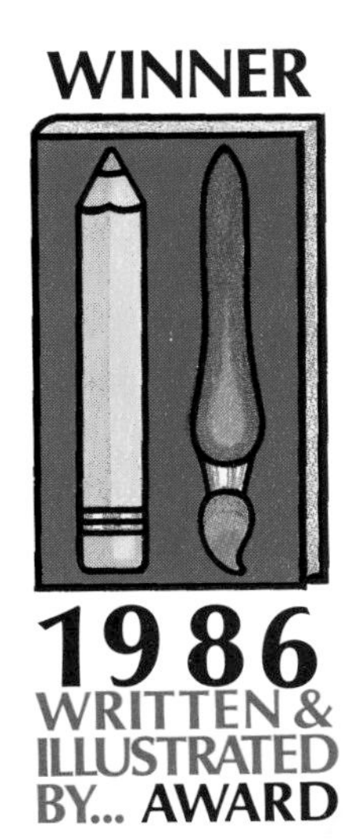

I was not surprised when ME AND MY VEGGIES, by Isaac Whitlatch, won THE 1986 NATIONAL WRITTEN & ILLUSTRATED BY... AWARDS CONTEST FOR STUDENTS in the age category of 10 to 13 years.

During the preliminary stages of judging, I could always tell when the judges were reading ME AND MY VEGGIES (and everyone else could too), because the reader would start chuckling, then interrupt the rest of us by reading an excerpt or two aloud.

Isaac's clever approach to satire comes to us in the tradition of Robert Benchley, Will Rogers, Dorothy Parker and Woody Allen. He too has the ability to see humor in the simple experiences of life and to project insights into the foibles of our human comedy. With deft insight and skill, Isaac nudges our funny bones.

The best writing transports readers into a complete environment. Isaac's narrative certainly does that — his assembly of words and observations carries us into the mind of a humorist. Soon the reader is seeing peas and carrots — not as nutritional — but as enemies to the taste buds of "an awesome kid."

Because Isaac is a child, it is too easy for the viewer to assume that his book illustrations represent the level of his drawing abilities. Having seen other examples of Isaac's drawings, I know better. For ME AND MY VEGGIES, however, Isaac has purposely used an outline-kid-style of drawing, because it is perfect for the mood and content of his delightful book.

The chuckles begin on the first page and don't stop until the last page of this fun-filled book.

— David Melton

Creative Coordinator
Landmark Editions, Inc.

I'm here to tell you — I am convinced there is a plot to destroy the health and sanity of the children of this world!

Behind every refrigerator door there lurks a variety of ghastly beasts. Three times a day, EVERY DAY, kids encounter the horrible little menaces. Splat! They are plopped onto our plates. Slop! They are scooped into our soup. They are hidden in Jello, ground into gravy, and heaped on our spoons.

And we are told, "Eat them, or else!"

What are kids to do? We either have to force them into our mouths, one by one, or gulp them down like hungry hippos before their horrid flavors invade our taste buds.

Enough is enough! It is time for children to revolt against daily attacks of these yucky, gooey, slimy things called VEGETABLES!

V
VEG
VEG
UEG
V
VEG
VEG

Like most kids, I was first introduced to VEGGIES in their baby food form. As my mother aimed spoonfuls of this pulverized pap at my unsuspecting mouth, she'd make silly sounds and say, "Buzz! Buzz! Here comes the airplane! Open the hangar door, and in it goes!"

What a dreadful trick to play on a baby! I'd try to spit out the awful stuff and let the gooey mess drool down my chin. You can bet one thing for sure — I soon learned when to open the hangar door and when to keep it shut!

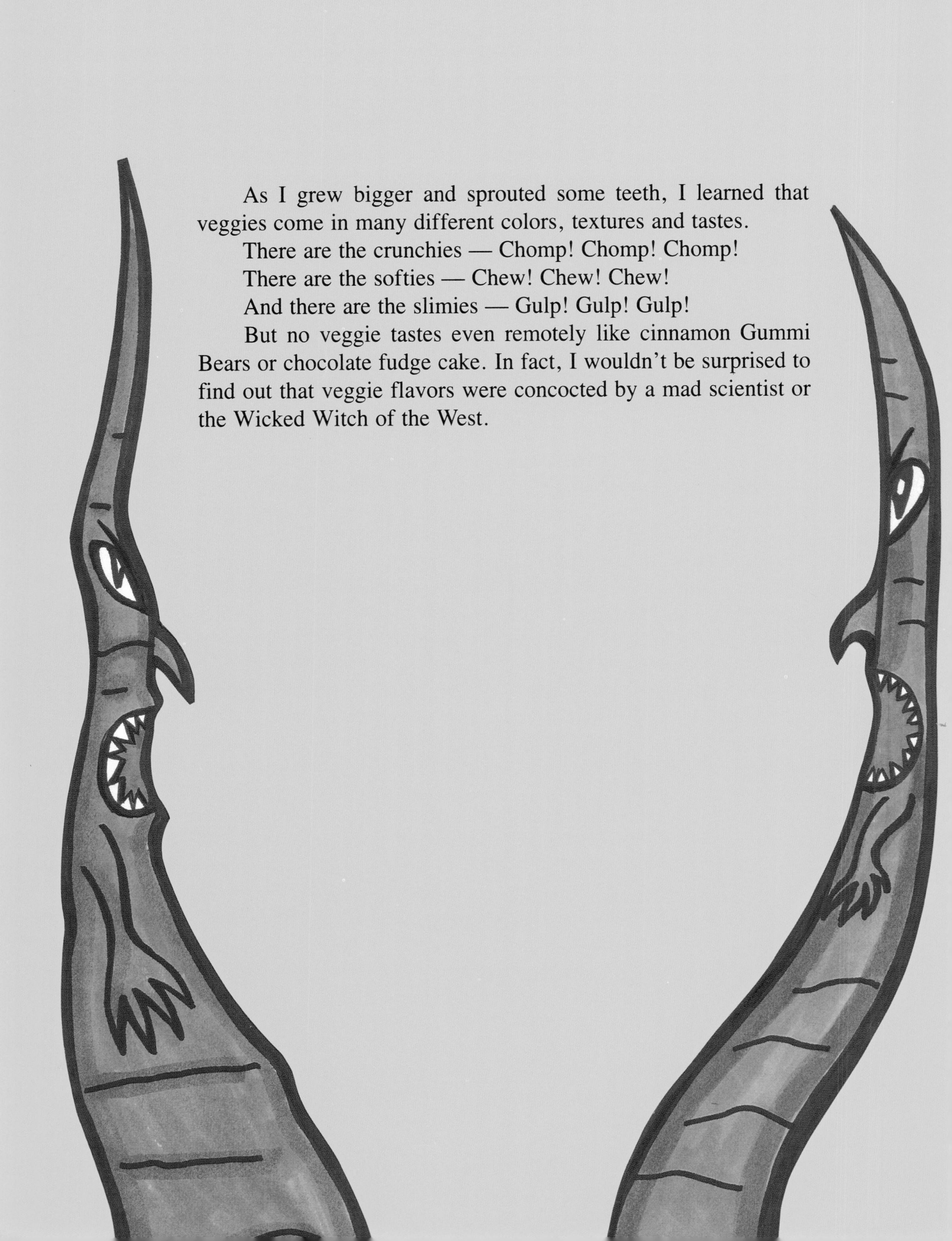

As I grew bigger and sprouted some teeth, I learned that veggies come in many different colors, textures and tastes.

There are the crunchies — Chomp! Chomp! Chomp!

There are the softies — Chew! Chew! Chew!

And there are the slimies — Gulp! Gulp! Gulp!

But no veggie tastes even remotely like cinnamon Gummi Bears or chocolate fudge cake. In fact, I wouldn't be surprised to find out that veggie flavors were concocted by a mad scientist or the Wicked Witch of the West.

My MOST UNWANTED LIST includes:

PEAS — Little green gremlins. When cooked, they're squishy. When frozen, they make terrific giant BBs.

CARROTS — Nothing but orange-colored roots.
"They're good for your eyes," my dad always tells me. Then he adds, "Did you ever see a rabbit wear glasses?"
Well, come to think of it, I never did, and I've never seen a rabbit wear a hearing aid either. Does that mean carrots are good for our ears too?

SQUASH — A perfect name! It SHOULD be squashed under the wheels of a speeding locomotive.

RADISHES — Very deceptive. Beautiful red coats on the outside, but awful-tasting white insides that will burn off the roof of your mouth in two seconds flat.

TOMATOES — Not a vegetable, but a fruit. Raw ones are often found sneaking around in salads.

BROCCOLI — A relatively new vegetable, developed in Europe, these creatures have now invaded American tables. An illegal alien that should definitely be deported.

SPINACH — A pile of lifeless, breathless, wilted goop!

LETTUCE — Let us NOT eat lettuce!

BRUSSELS SPROUTS — Nothing but bitter dwarf cabbages!

MUSHROOMS — A fungus too horrible to describe!

ASPARAGUS — Long, slick spears with a taste that crumbles your teeth and withers your tongue! Of all the veggies, it's the most treacherous. I'm convinced that asparagus will stunt my growth and shorten my life.

However, I don't care what anyone says — French fries are NOT vegetables! And popcorn doesn't count either.

Kids, I warn you, our vegetable enemies outnumber and outflank us on all sides. Through devious maneuvers and masses of propaganda, the veggies brainwash adults — especially mothers!

My mother is certainly on their side. She is a health food fanatic. She loves broccoli...and carrots...and green beans... and peas...and asparagus as much as Baskin-Robbins German chocolate ice cream.

Mom's a pro when it comes to preparing veggies. She cuts 'em, chops 'em, shreds 'em, peels 'em, boils 'em, bakes 'em, fries 'em, and even carrot-cakes 'em. And just when you think it's safe to go back into the kitchen, she invents a new recipe for cooking more of the little monsters.

I've finally decided that it's ALL-OUT WAR! To combat Colonel Cauliflower and his leafy armies, I have developed some well-planned maneuvers. My defensive strategies have two main divisions — SCATTERING and HIDING.

SCATTERING

Scattering is an ancient art, but I have perfected it. First, I divide the pile of veggies into fourths. Then I carefully move one pea, or one green bean, or one small carrot to each corner of the plate.

Believe me, it's a real trick to find a corner on a round plate.

My ultimate goal is to have only a few pieces touching each other. When the piles are scattered enough, they give the illusion that most of the veggies have been eaten. If Mom thinks I've ALMOST cleaned my plate — I've won!

HIDING

Hiding requires the art of skillful camouflage. It is a simple fact — for hiding veggies, a glass of milk is better than a glass of iced tea. Peas show up in iced tea, but a few inches of milk will cover a couple of brussels sprouts, or two carrot sticks, or a spoonful of peas.

I love paper napkins. When fortunate enough to get one, I pretend to blow my nose, while I'm really spitting unwanted veggies into the napkin folds. I never try that with a cloth napkin, because when Mom does the laundry, I know she will discover my little garbage dump.

When the phone rings during mealtimes, it's a great stroke of luck! While my mom's away from the table, if I work quickly, I can camouflage the last of my veggies under chicken bones, crusts of bread or aluminum foil wrap from a baked potato.

INFESTED
water

I prefer to wear shirts that button down the front, because I sometimes push larger pieces of veggies inside. Green beans go under my baseball cap. But once I got caught with a green bean dangling behind my left ear. So I now check and double-check to make sure everything is tucked neatly away out of sight.

Elastic is great! Around sock tops, it provides the perfect place to hide peas and other small bits. When I have pencils in my back pocket, a few asparagus spears blend in unnoticed. And my Levis definitely give veggies the 501 blues.

My mom kept saying, "Clean your plate, because there are starving children in India."

So one day, when Mom wasn't looking, I scraped the leftover veggies from my plate into a box and mailed it to "The Starving Children of India."

I didn't know you had to put stamps on gifts. Just my luck! Three days later, the box came back.

When Mom opened the box, she almost fainted. Each of the veggies was covered with three inches of green mold.

It makes one wonder: If veggies grow three inches of green mold in only three days in a box, just think what they might do in your stomach.

Besides SCATTERING and HIDING, I also have some special SECRET WEAPONS — my cat, my little brother, and the toilet stool. And as a last resort there is the Boston fern in the corner of the dining room.

SECRET WEAPON NUMBER 1 —

MY CAT

My cat, Manassas, is an exotic breed called Abyssinian. He is a rarity for sure. His personality is more like a dog's than any cat's I've ever seen.

Manassas looks like a mountain lion and eats like an alligator. Once he chomped up two chameleons, so I definitely know he likes green wiggly things.

When Mom isn't looking, I shake a green bean in front of that cat's nose, and Chomp! — it's a goner! When I toss a carrot strip into the air, the feline flash will chase it, catch it, kill it, and bury it with great pride in his scented kitty litter.

I highly recommend that the reader buy an Abyssinian cat...TODAY!

SECRET WEAPON NUMBER 2 —

MY LITTLE BROTHER

My little brother, Luke, is weird. He actually likes rabbit food — peas, corn, broccoli, and even squash. Yuck!

Fortunately Luke will also eat MY veggies, that is, for a price. Carrots require only a cheap bribe — maybe a nickel. Green beans take a little more persuasion — perhaps a couple of candy bars. But asparagus sometimes costs me a whole week's allowance.

However, due to clever tactics on my part, my unsuspecting brother frequently eats a few of my foliage foes free. Using my table utensils as military hardware, I secretly flip my veggies onto his plate. During these military maneuvers, I classify the common spoon as a HIGH VELOCITY PEA LAUNCHER.

$
Snake

SECRET WEAPON NUMBER 3 —

THE BOSTON FERN

The dining room battlefield has its own burial ground — in the pot of the Boston fern. As I see it, once a veggie is cut off from its mother source, it's obviously dead, or at least dying. So I figure the only humane thing to do is to bury it before it begins to stink. Right? Right!

My motto is: *A bean without its pod needs a peaceful resting place for its bod.*

I've given many dying veteran veggies a full funeral and decent burial in the rich and ever-improving soil of the Boston fern. I tell my veggies, "If you will cooperate and become good mulch, you will go straight to that "Big Garden in the Sky!"

SECRET WEAPON NUMBER 4 —

THE TOILET STOOL

To get captured veggies from the table to the toilet stool, without getting caught, takes super tactics on my part. I start by eating slowly, so I'll be the last one to leave the table. Then while everyone else is zonked in on television, I make my move. Taking care not to attract attention, I stuff my mouth full of veggies and slowly make my way toward the *you know what*.

My special commando training keeps me from dropping any secrets held in my mouth. I have even learned how to give my name, rank and serial number while holding a whole serving of peas under my tongue. Of course, I try to answer other questions with a simple "Uh-huh" or "Huh-uh," while taking my veggies on their final trip to the "Grand Flush."

There are times when I am not able to use my SECRET WEAPONS — my little brother stubbornly refuses all bribes; my cat's outside; my dad's in the bathroom; or Mom's watering the Boston fern.

That's when I'm forced to use the old DISSECT-AND-SWALLOW TECHNIQUE. At these times, I have no choice but to cut my vitamin-filled veggies into teeny, tiny pieces and swallow them with a gulp of water — like when I'm taking a pill. That way, I don't have to taste them. However, if I down too many, too quickly, my stomach soon feels like I've swallowed a whole watermelon, seeds and all!

BURP
THE Pits

My dad reminds me that our Great Aunt Harriet was a vegetarian — she ate only vegetables. Of course, she died before I was born, which I think proves my point — if you eat too many veggies, they'll kill you!

I have thought it over and I have concluded: The only green things I like are Gatorade and dollar bills.

Believe it or not, I found a verse in the BIBLE that says:
"...he who is weak eats vegetables."
— Romans 14:2, New King James Version

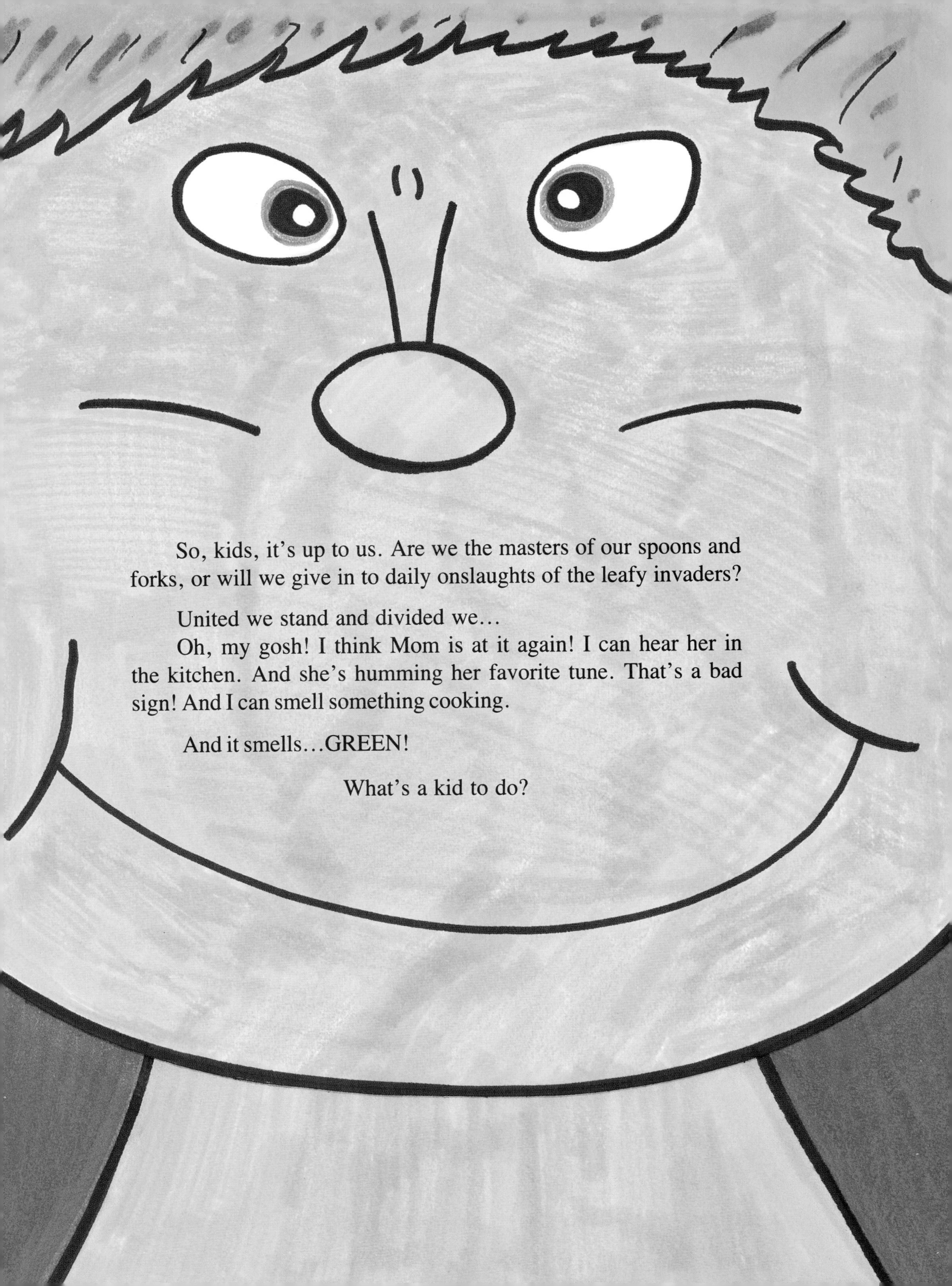

So, kids, it's up to us. Are we the masters of our spoons and forks, or will we give in to daily onslaughts of the leafy invaders?

United we stand and divided we…

Oh, my gosh! I think Mom is at it again! I can hear her in the kitchen. And she's humming her favorite tune. That's a bad sign! And I can smell something cooking.

And it smells…GREEN!

What's a kid to do?

THE NATIONAL WRITTEN & ILLUSTRATED

— THE 1989 NATIONAL AWARD WINNING BOOKS —

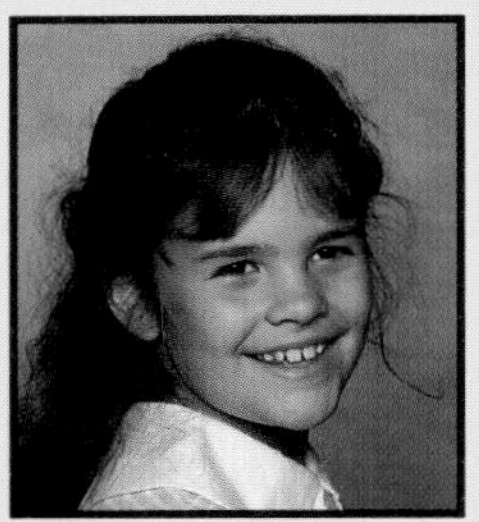

Lauren Peters
age 7

the Legend of SIR MIGUEL
written and illustrated by MICHAEL CAIN

WE ARE A THUNDERSTORM
written and photographed by
amity gaige

Michael Cain
age 11

—THE 1987 NATIONAL AWARD WINNING BOOKS—

Who Owns The Sun?
-written & illustrated by-
STACY CHBOSKY

Amity Gaige
age 16

Dennis Vollmer
age 6

—THE 1989 GOLD AWARD WINNERS—

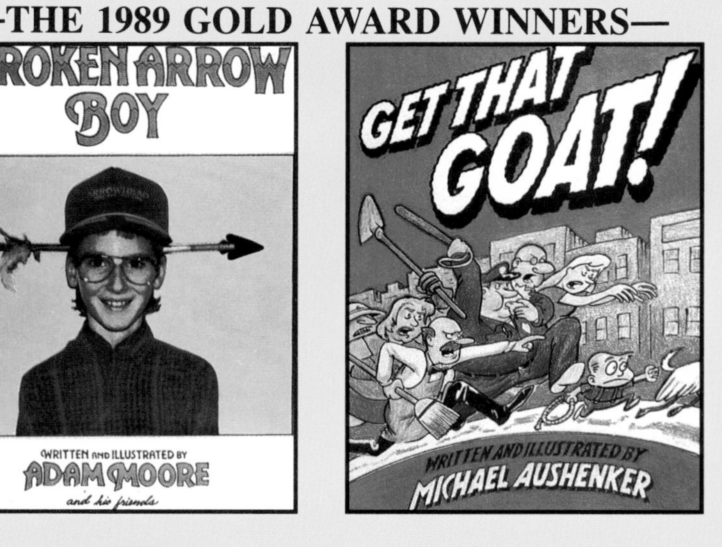

GET THAT GOAT!
WRITTEN AND ILLUSTRATED BY
MICHAEL AUSHENKER

Lisa Gross
age 12

Stacy Chbosky
age 14

Adam Moore
age 9

Michael Aushenker
age 19

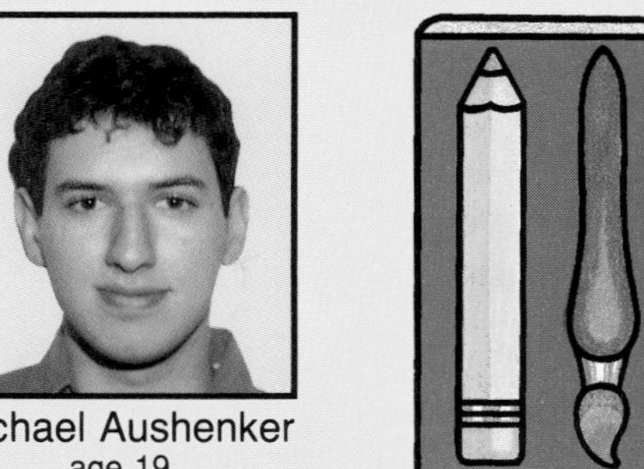

Students' Winning Books Motivate and Inspire

Each year it is Landmark's pleasure to publish the winning books of The National Written & Illustrated By... Awards Contest For Students. These are important books because they supply such positive motivation and inspiration for other talented students to write and illustrate books too!

Students of All Ages Love the Winning Books

Students of all ages enjoy reading these fascinating books created by our young author/illustrators. When students see the beautiful books, printed in full color and handsomely bound in hardback covers, they, too, will become excited about writing and illustrating books and eager to enter them in the Contest.

Enter Your Book In the Next Contest

If you are 6 to 19 years of age, you may enter the Contest too. Perhaps your book may be one of the next winners and you will become a published author and illustrator too.

BY... AWARDS CONTEST FOR STUDENTS

— THE 1988 NATIONAL AWARD WINNING BOOKS —

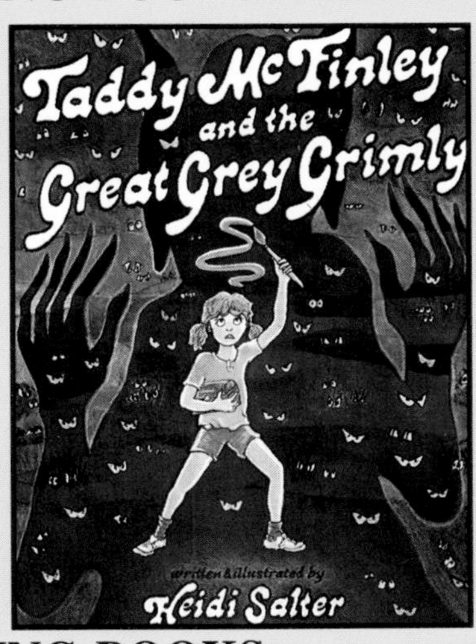

Leslie Ann MacKeen
age 9

Elizabeth Haidle
age 13

—THE 1986 NATIONAL AWARD WINNING BOOKS—

WORLD WAR WON
WRITTEN & ILLUSTRATED BY
DAV PILKEY

Heidi Salter
age 19

— THE 1985 GOLD AWARD WINNERS —

Winners Receive Contracts, Royalties and Scholarships

The National Written & Illustrated by... Contest Is an Annual Event! There is no entry fee! The winners receive publishing contracts, royalties on the sale of their books, and all-expense-paid trips to our offices in Kansas City, Missouri, where professional editors and art directors assist them in preparing their final manuscripts and illustrations for publication.

Winning Students Receive Scholarships Too! The R.D. and Joan Dale Hubbard Foundation will award a total of $30,000 in scholarship certificates to the winners and the four runners-up in all three age categories. Each winner receives a $5,000 scholarship; those in Second Place are awarded a $2,000 scholarship; and those in Third, Fourth, and Fifth Places receive a $1,000 scholarship.

To obtain Contest Rules, send a self-addressed, stamped, business-size envelope to: THE NATIONAL WRITTEN & ILLUSTRATED BY... AWARDS CONTEST FOR STUDENTS, Landmark Editions, Inc., P.O. Box 4469, Kansas City, MO 64127.

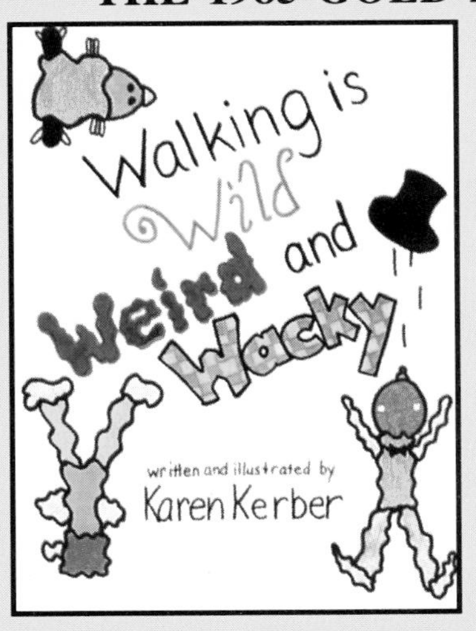

Amy Hagstrom
age 9

Isaac Whitlatch
age 11

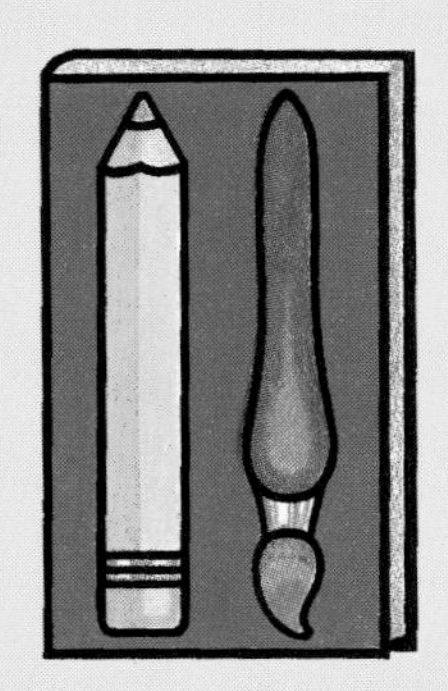

Karen Kerber
age 12

David McAdoo
age 14

Dav Pilkey
age 19

Valuable Resources for Teachers and Librarians Proven Successful in Thousands of Schools

Written & Illustrated by...

by David Melton

This highly acclaimed teacher's manual offers classroom-proven, step by-step instruction in all aspects of teaching students how to write, illustrat and bind original books. Contains suggested lesson plans and more thar 200 illustrations. Loaded with information and positive approaches tha really work.

. . . an exceptional book! Just browsing through it stimulates excitement for writing.
Joyce E. Juntune, Executive Director
National Association for Gifted Children

The results are dazzling!
Children's Book Review Service, Inc.

A "how to" book that really works!
Judy O'Brien, Teacher

This book should be in every class room.
Tacoma Review Service

96 Pages
Over 100 Illustration
Softcover
ISBN 0-933849-00-1

How To Capture Live Authors
and Bring Them to Your Schools

by David Melton

If you want to initiate and improve your Young Authors' Days, Children's Literature Festivals and author-related programs — this book is for you!

. . . covers everything from initiating "The Literary Event of the School Year" to estimating final costs. The Checklists are invaluable. Teachers, librarians and parents will want to read this book from cover to cover.
School Library Journal

Thanks to this wonderful book, our first Visiting Author Program worked perfectly. Thank you! Thank you!
Trudie Garrett, School Librarian

This book is full of down-to-earth information.
Booklist

A unique book — practical and entertaining.
The Missouri Reader
International Reading Association

96 Pages
Illustrated
Softcover
ISBN 0-933849-03-6

We believe the creativity of students is important. We are delighted that our unique materials and programs provide hundreds of thousands of students with opportunities to create original books.

We are accustomed to receiving the friendliest and most enthusiastic letters from teachers, librarians and parents from literally all parts of the world. We love to hear about successful author-related programs and how the lives of students and teachers have been affected in positive ways.

We treasure your acceptance and your friendship. Your continued support is so important in the expansion of our programs to offer increased creative opportunities for students.

Thank you.

LANDMARK EDITIONS, INC.
P.O. Box 4469 • 1402 Kansas Avenue • Kansas City, Missouri 64127
(816) 241-4919